Energy

Lisa Trumbauer

Contents

Why We Need Energy

All living things need **energy.** Trees need energy to grow.

Birds need energy to grow and fly.

You need energy to move, learn, and play! Where does all this energy come from?

Energy from the Sun

One **source** of energy is the sun.
The sun is a huge glowing ball of gases.

The sun gives us light and heat. You use this energy every day.

The sun's light and heat are forms of energy. All the living things around you use the sun's energy.

Plants use sunlight to make their own food.

The food gives them energy to grow.

Many animals eat plants for food.
This food gives the animals energy.

People eat plants and animals for food. The food people eat gives them energy.

Energy from Earth

Other sources of energy are found in the earth. Natural gas, oil, and coal lie deep underground.

drilling for oil

coal

We burn natural gas, oil, and coal for energy.

This stove burns natural gas.
The heat cooks our food.

This cabin is heated by a furnace.
The furnace burns oil.

Gasoline is made from oil. Most cars burn gasoline to make them go.

Electrical Energy

Most **power plants** burn coal or oil to make **electricity,** which is a form of energy.

coal-burning power plant

The electricity is carried through power lines to our homes.

We use electricity to light our homes and to run our computers, refrigerators, and other **appliances**.

Water, wind, and the sun can also be used to make electricity.

Dams use energy from flowing water to make electricity.

Wind turbines use energy from the wind to make electricity.

Solar panels collect the sun's energy to make electricity.

Look around you.
Energy is everywhere.

What forms
of energy
make these
things work?

Glossary

appliance (uh-PLYE-uhns): a machine designed for a specific job

dam (DAM): a barrier built across a river or stream to control the flow of water

electricity (ih-lek-TRIH-suh-tee): a form of energy that gives us light, heat, and sound

energy (E-nur-jee): the ability to do work

power plant (POW-ur PLANT): a place where electricity is generated

solar panel (SOH-lur-PA-nul): an object that collects the sun's energy

source (SORS): the place something comes from

wind turbines (WIND TUR-buns): machines that use wind power to make electricity